P9-DDU-321

GOODNIGHT iPAD

a Parody for the next generation

by Ann Droyd

blue rider press

a member of Penguin Group (USA) Inc.

New York

LICENSE AGREEMENT

blue rider press

BLUE RIDER PRESS
Published by the Penguin Group
Penguin Group (USA) Inc., 375 Hudson Street, New York, New York 10014, USA
Penguin Group (Canada), 90 Eglinton Avenue East, Suite 700, Toronto, Ontario M4P 2Y3,
Canada (a division of Pearson Penguin Canada Inc.)
Penguin Books Ltd, 80 Strand, London WC2R 0RL, England
Penguin Ireland, 25 St Stephen's Green, Dublin 2, Ireland (a division of Penguin Books Ltd)
Penguin Group (Australia), 250 Camberwell Road, Camberwell, Victoria 3124, Australia (a division of Pearson Australia Group Pty Ltd)
Penguin Books India Pvt Ltd, 11 Community Centre, Panchsheel Park, New Delhi–110 017, India
Penguin Group (NZ), 67 Apollo Drive, Rosedale, North Shore 0632, New Zealand (a division of Pearson New Zealand Ltd)
Penguin Books (South Africa) (Pty) Ltd, 24 Sturdee Avenue, Rosebank, Johannesburg 2196, South Africa

Penguin Books Ltd, Registered Offices: 80 Strand, London WC2R 0RL, England

Copyright © 2011 by David Milgrim
All rights reserved. No part of this book may be reproduced, scanned, or
distributed in any printed or electronic form without permission.
Please do not participate in or encourage piracy of copyrighted materials
in violation of the author's rights. Purchase only authorized editions.
Published simultaneously in Canada

ISBN 978-0-399-15856-8

Printed in the United States of America but not in Cupertino, California
20 19 18 17 16 15 14 13 12 11

Book design by Michael Nelson

This is a work of fiction. Names, characters, places, and incidents either are the product of the author's imagination or are used fictitiously, and any resemblance to actual persons, living or dead, businesses, companies, events, or locales is entirely coincidental.

Don't bother reading this. No one does. Just scroll to the bottom and click.

I have read the License Agreement Terms and Conditions **I AGREE**

I have not read the License Agreement Terms and Conditions **I AGREE anyway**

For super-agent Brenda Bowen, who made
this book possible in every way

And for everyone
who is as hopelessly
plugged in as I am

In the bright buzzing room
There was an iPad
And a kid playing Doom
And a screensaver of—

A bird launching over the moon

There were three little Nooks
With ten thousand books

And a huge LCD
Wi-Fi HDTV

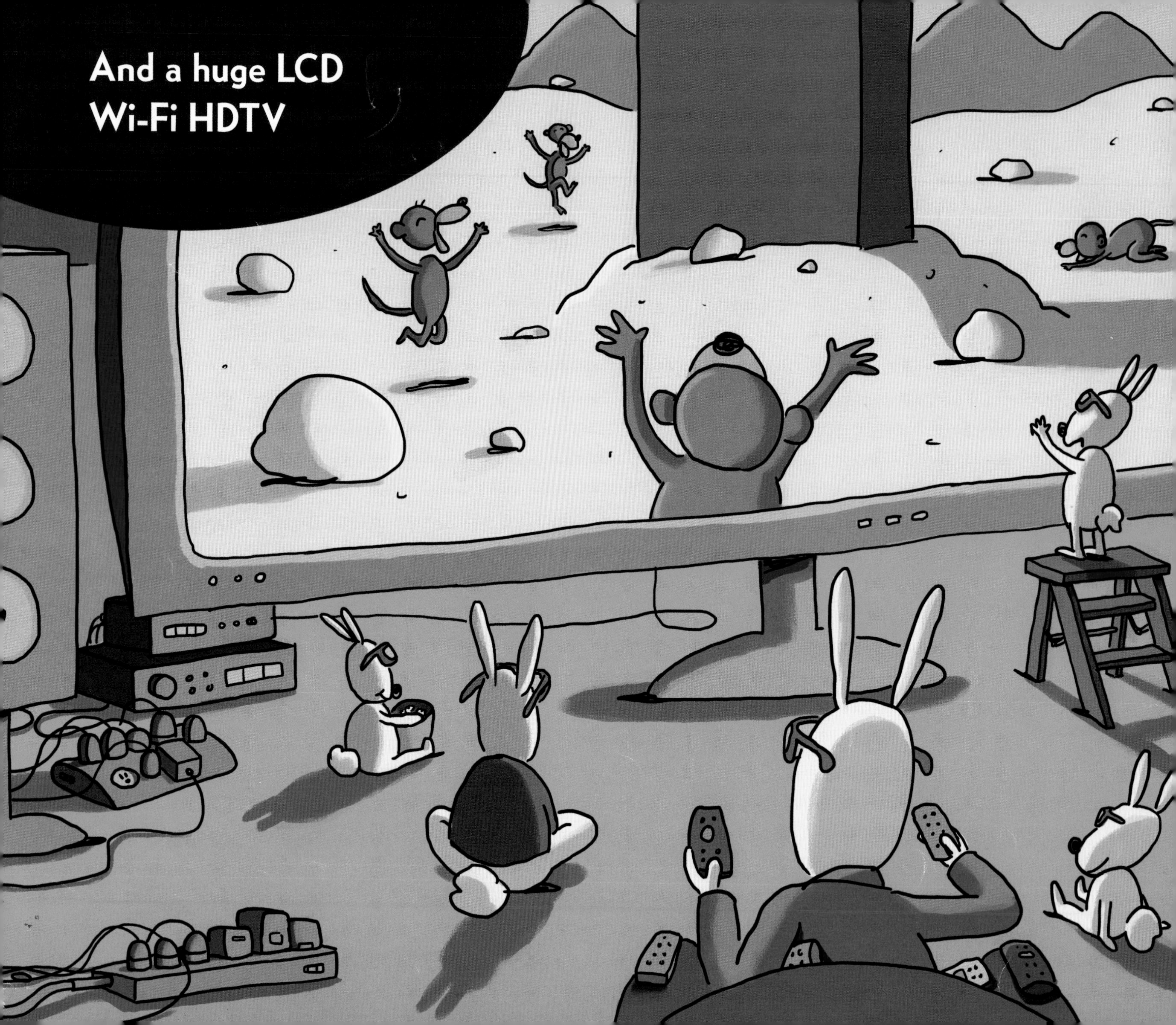

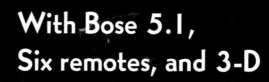

With Bose 5.1,
Six remotes, and 3-D

And a BlackBerry ringing
With Eminem singing

And a new Facebook friend
And texts with no end

And a viral clip of a cat doing flips

And the bings, bongs, and beeps
Of e-mails and tweets

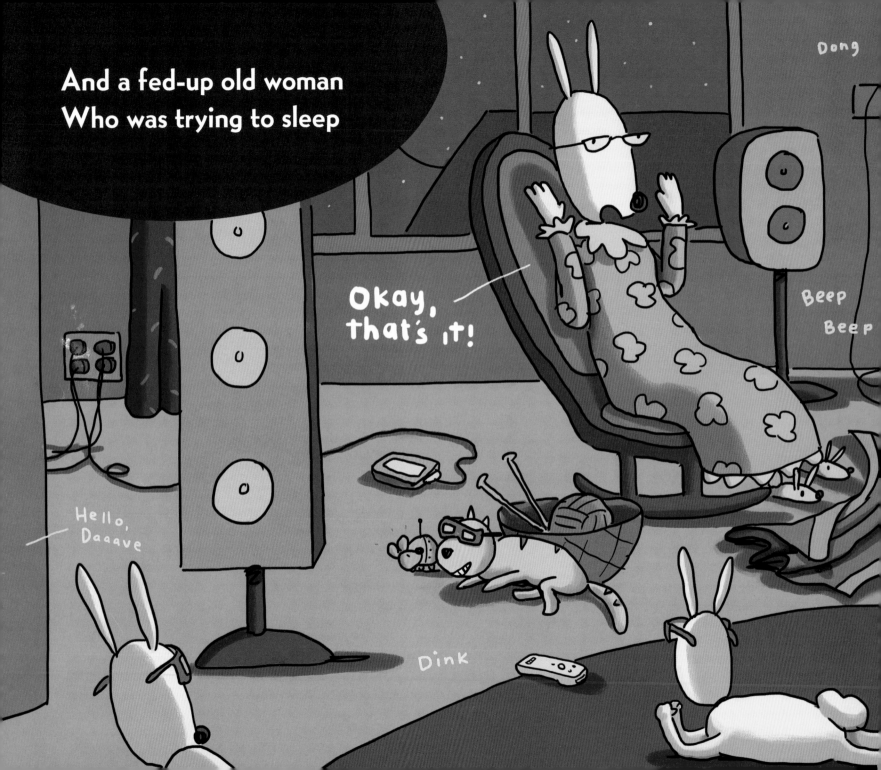

Goodnight Doom

Goodnight bird
Launching over the moon

Goodnight Nooks
And digital books

Goodnight Eminem
Goodnight Facebook friend

Goodnight LOLs
Goodnight MP3s

Goodnight LCD Wi-Fi HDTV

Goodnight remotes
And Netflix streams,
Androids, apps,
And glowing screens

Okay,
everyone...

Bedtime!

Goodnight plugs
And power lights
That guide us to pee
In the darkness of night

Goodnight buzzing
Goodnight beeps

Goodnight everybody
Who should be asleep

Goodnight pop stars

Goodnight MacBook Air

Goodnight gadgets everywhere